Interagency Steering Committee on Radiation Standards

2002 Annual Report

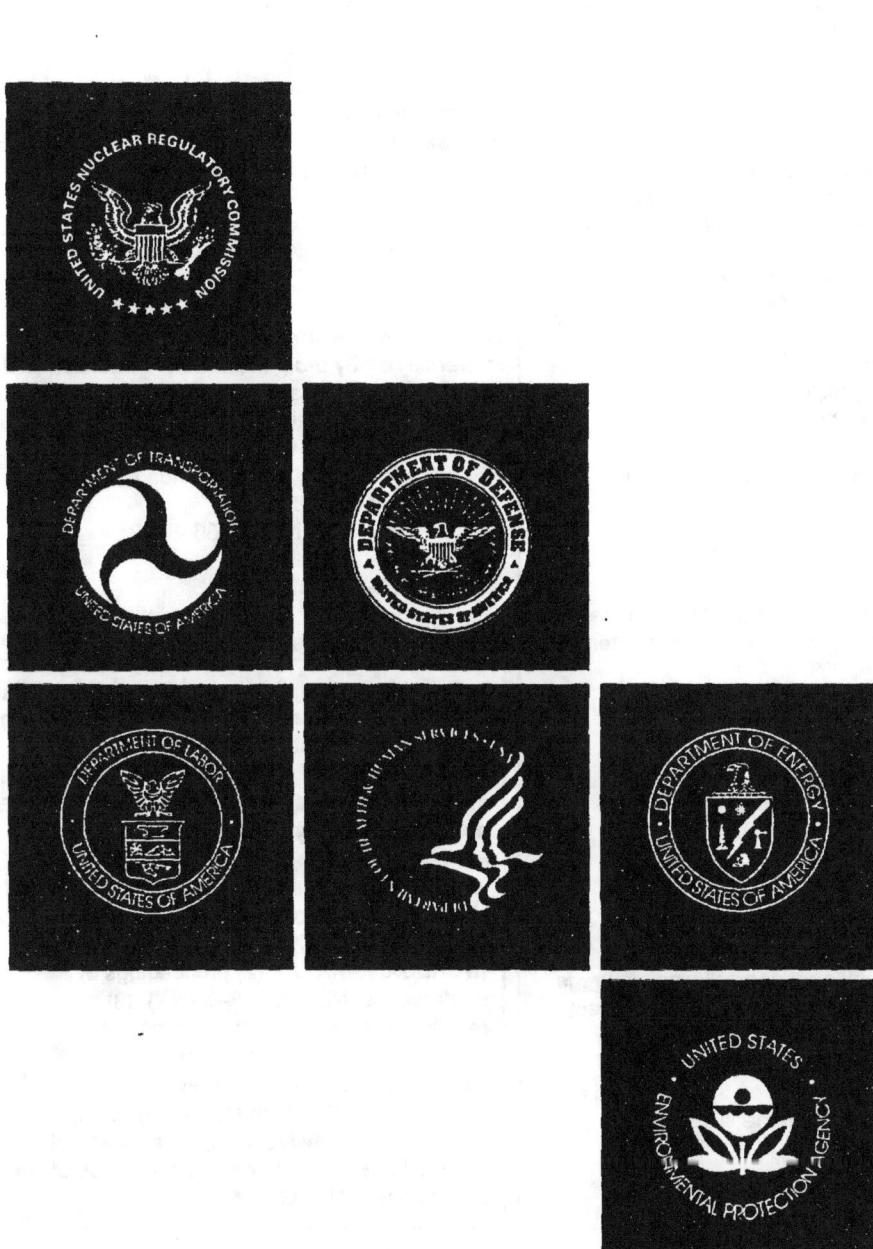

NUREG-1707, Vol. 5
ISCORS-03-01

AVAILABILITY OF REFERENCE MATERIALS
IN NRC PUBLICATIONS

INTERAGENCY STEERING COMMITTEE ON
RADIATION STANDARDS

2002 ANNUAL REPORT

TABLE OF CONTENTS

FOREWORD

The Interagency Steering Committee on Radiation Standards (ISCORS) prepared this annual report for ISCORS member agencies to document ISCORS' activities and accomplishments in 2002, and plans for 2003. We wish to thank the ISCORS members for their participation and contributions to the many topics discussed over the past year regarding radiation issues important to the public; federal, state, and local agencies; and national and international communities. In addition, we extend our compliments to all the subcommittee chairs and their members for significant support and accomplishments within the past year on numerous radiation issues, and for developing useful information, examples of which are attached in this report and on the ISCORS website (www.iscors.org). The subcommittees have outlined an aggressive list of challenges for 2003 and the full committee looks forward to receiving subcommittee recommendations on the specific topics each is addressing.

The full ISCORS met in January, April, July, and October 2002. The July 2002 meeting was open to the public for observation. At each full committee meeting, the subcommittees reported on their yearly activities and progress. The full committee also reviewed and received focused presentations on a wide variety of special topics, some of which included the following:

- Revision of the Protective Action Guides (PAGs)
- Memorandum of Understanding (MOU) between the NRC and EPA, and
- Updates on the activities of the International Commission on Radiation Standards (ICRP), the National Commission on Radiation Standards (NCRP), and the National Academies of Science (NAS).

Over the past few years, the full ISCORS committee has established a strong intragovernmental working relationship that benefits each member agency and significantly aids in identifying topics of interest to each agency. For both member representatives and the public, the ISCORS process and accomplishments are made visible by the annual report. We appreciate comments on this annual report and will use them to consider how ISCORS might better carry out its responsibilities. You may send comments on the report to Mr. James E. Kennedy (U.S. NRC, Mail Stop T-7J8, Washington, DC 20555), or Ms. Kathryn A. Klawiter (U.S. EPA, Office of Radiation and Indoor Air (6608J), 1200 Pennsylvania Avenue, NW, Washington, DC 20460).

ISCORS Co-chairs:

Mr. John T. Greeves, Director
Division of Waste Management, Mail Stop
   T-7J8
Office of Nuclear Material Safety and
   Safeguards
U.S. Nuclear Regulatory Commission
Washington, DC 20555

Mr. Frank Marcinowski, Director
Radiation Protection Division
Office of Radiation and Indoor Air (6608J)
U.S. Environmental Protection Agency
1200 Pennsylvania Avenue, NW
Washington, DC 20460

HISTORICAL INFORMATION

The Interagency Steering Committee on Radiation Standards (ISCORS) was formed in response to October 27, 1994, letters from Senator John Glenn to the U. S. Nuclear Regulatory Commission (NRC) the U.S. Environmental Protection Agency (EPA); and the Office of Science and Technology Policy (OSTP). In this letter, Senator Glenn charged the EPA and the NRC, in coordination with the Committee on Interagency Radiation Research and Policy Coordination (CIRRPC), to develop a plan for a "path forward" to address the inconsistencies, gaps, and overlaps in current radiation protection standards. ISCORS is one of the committees OSTP recommended for achieving the goals of the now defunct CIRRPC. The objectives of the ISCORS include the following: (1) facilitating a consensus on acceptable levels of radiation risk to the public and workers; (2) promoting consistent risk-assessment and risk-management approaches in setting and implementing standards for occupational and public protection from ionizing radiation; (3) promoting completeness and coherence of federal standards for radiation protection; and (4) identifying interagency issues and coordinating their resolution.

Since its inception, the NRC and the EPA have co-chaired the ISCORS. The current co-chairs are John T. Greeves, NRC, and Frank Marcinowski, EPA. In addition to the NRC and the EPA, ISCORS membership also includes senior managers from the U.S. Department of Defense; the U.S. Department of Energy (DOE); the U.S. Department of Labor Occupational Safety and Health Administration (OSHA); the U.S. Department of Transportation (DOT); and the U.S. Department of Health and Human Services (DHHS). Representatives of the Office of Management and Budget (OMB), OSTP, and various states are observers at meetings. Committee meetings involve pre-decisional intragovernmental discussions and, as such, are not normally open for observation by members of the public or media. However, summary meeting notes are available at the ISCORS website, (www.ISCORS.org). The Committee does not act as a decision-making body. Instead, it provides recommendations and summaries of its activities on specific issues to both the heads of member agencies, and to OMB and OSTP, via an annual report. The Committee meets approximately once each calendar quarter. It held its first meeting on April 5, 1995.

The full ISCORS committee has established subcommittees, as needed, to conduct its technical work (e.g., to address specific issues of concern or significant interest). ISCORS has formed the following subcommittees: Clean-up; Mixed Waste; Recycle; Risk Harmonization; Sewage Sludge; Naturally Occurring Radioactive Materials (NORM); and Federal Guidance. The subcommittee activity section of this report summarizes each subcommittee's activities. No new subcommittees were formed in 2002.

SUMMARY OF SUBCOMMITTEE ACCOMPLISHMENTS IN 2002 AND PLANNED ACTIVITIES FOR 2003

Clean-up Subcommittee Highlights

Subcommittee Chair: Anthony Wolbarst, EPA (202-564-9392)

The Clean-up Subcommittee serves as a forum for exchanging information and promoting partnerships addressing radiation clean-up standards and guidance.

Accomplishments in Calender Year 2002 (CY 2002)

1. Initiation of a New Project. The subcommittee members compiled a list of new projects of potential interest to their respective agencies, and decided to focus energy and resources initially on one of them. This first project involves the creation of an organized, searchable compilation of the references, compendiums, and databases of parameters used by environmental fate and transport models in dose and risk assessments. This compilation, entitled the *ISCORS Catalog of Existing Sources of Information on Parameters Used in Pathway Modeling for Environmental Cleanup of Sites Contaminated with Radioactive Materials*, will provide modelers with an efficient means of locating extant sources of parameter values and distributions. It will include available information on the QA/QC used by the sources in data acquisition, and listings of any parameter default values recommended by agencies.

   The catalog is being prepared, under the direction of the Subcommittee, through Inter-Agency Agreements with the Department of Energy's Argonne National Laboratory (ANL). The project is being carried out in two phases. The first phase, scheduled for FY 2003, consists of six tasks: identify, collect, compile information on sources of parameter information; develop an indexing system; create an initial, restricted website for development work; check out and improve the indexing system by applying it to the initial parameter-source data; demonstrate the system operation; and analyze the feedback from the demonstration and from outside reviewers, and prepare a plan for testing the system and website. ANL has begun work on these tasks. The second and final phase, or completion phase, of the project is scheduled for 2003 and 2004.

Activities Planned for CY 2003:

1. Completion of the *Parameter-Source Catalogue*. The completion phase of development of the *Parameter-Source Catalogue* involves five tasks: Respond to feedback following the system operation demonstration and also from outside reviewers; finalize the website (which must be efficient, well-documented and user-friendly, and uncluttered and aesthetically pleasing); test the system, including the website; gather, enter, and index more parameter-source data; and, after obtaining final ISCORS approval, release the website to the public. Issues for the Subcommittee to consider are the development and implementation of plans to bring the catalog website to the attention of large numbers of potential users, to provide technical support and a channel for feed-back from users, and to provide for the routine updating and expansion of its contents.

Possible Additional Projects for 2003:

1. A catalog of existing information on federal and state agency policies and practices regarding environmental pathway modeling, including: rulemaking *vs.* screening *vs.* compliance demonstration modeling; model and data QA/QC requirements; deterministic *vs.* probabilistic modeling; demonstration of compliance with probabilistic and other dose/risk limits; conservative *vs.* best estimate parameter values; determination of the need for site-specific data; default values and distributions in the absence of site-specific data; peak dose versus lifetime risk; dealing with background radiation; combined chemical and radiation risks; peer review; expert judgement; and other policy-related topics.

2. A descriptive atlas or catalog of generic and site-specific modeling scenarios currently in existence. The catalog will describe general approaches employed in the development of scenarios, and provide agencies and stakeholders with example scenarios to serve as templates for use in cleanup and other activities. Generic scenarios would include those used in the development of the NRC and EPA cleanup rules, or dealing with sewage sludge and TENORM. Site-specific scenarios would include examples from NRC license termination, EPA Superfund remediation, and other agency programs.

3. A two-part descriptive catalog that focuses primarily on worker safety. The first part will include existing information and lessons learned regarding radiation hazards at non-radiation facilities, including: oil production and transport (waste water and oil scale), phosphate fertilizer, metal mining and processing waste, geothermal energy production wastes, water treatment plant residues, and landfills receiving such waste streams. It will also cover agency guidance on radiation landfills receiving such waste. The second part of the catalog will present information on potential chemical hazards at radiation sites, including methods of detection and related technical issues.

Mixed Waste Subcommittee Highlights

Subcommittee Chair: Gus Vazquez, DOE (202-586-7629)

The Mixed Waste Subcommittee is a forum for exchanging information on hazardous-radioactive waste (mixed waste).

Accomplishments in Calender Year 2002 (CY 2002)

1.   Individual Agency Initiatives. Individual agency members pursued agency-specific mixed waste initiatives that were not yet ready for Subcommittee review. DOE worked on advancing the Radiological Control Criteria (RCC) waste management approach for most of 2002. Timely updates of DOE activities were provided to the Subcommittee members, including summaries of interactions with State officials from CRCPD and ASTSWMO. EPA and NRC began moving forward their Low Activity Mixed Waste (LAMW) initiative in 2002. DOE and State representatives agreed not to pursue parallel development of RCC, but to refocus RCC efforts to support the new LAMW initiative, because it will be based on a Federal rulemaking. DOE provided a briefing to EPA-ORIA on its RCC work in October 2002. In December 2002, DOE received an official letter from the CRCPD's E-5 Committee chairperson regarding the Committee's general agreement and concurrence with the technical work underlying RCC, and recommended that this work be used to support LAMW.

2.   Subcommittee Coordination. The Subcommittee did not meet, in large part because of agency member schedule delays beyond its control, and was, therefore, unable to meet all of its expected four 2002 goals. The Subcommittee continued to coordinate via e-mail and telephone.

Activities Planned for CY 2003:

1.   Review and provide input to the Low Activity Mixed Waste (LAMW) initiative.

2.   Review and provide input to the Low Activity Waste (LAW) initiative.

3.   Provide a continuing forum for information exchange and coordination between the agencies on matters related to mixed waste.

Recycle Subcommittee Highlights

Subcommittee Chair: Robert A. Meck, NRC   (301-415-6205)

The Recycle Subcommittee reviews issues on radiological control of materials.

Accomplishments in CY 2002:

1.   Federal Agency Activities --The subcommittee kept well informed of federal agency activities and international developments. The Programmatic Environmental Impact Statement process of DOE on the disposition of metals with very low levels of associated radioactivity is in preparation. Active information exchanges informed other agencies on the NRC's efforts to consider alternatives for disposition of solid materials with very low levels of associated radioactivity. These exchanges included the results of the National Academies' review of the subject.

2.   International Developments --The International Atomic Energy Agency (IAEA) shifted from deriving concentrations for clearance to proposing concentrations that would define the scope of regulations then to concentrations not requiring regulation for radiation protection purposes. The subcommittee used its interagency representation to develop a U.S. position on the draft IAEA Safety Guide, "Radionuclide Content in Commodities not requiring Regulation for Purposes of Radiation Protection," and reviewed a revised draft.

3.   Orphan Source Control --The subcommittee kept informed on the CRCPD and Federal agencies' program to identify and control orphan sources. To date five States have signed agreements and in three States agreement are being processed. Using 2 grapple mounted radiation detectors, EPA monitored over 74,000 tons of scrap metal entering the US through the Port of New Orleans. No radioactive materials were found in the 4 shipments monitored. The project is being expanded to the Port of Charleston, S.C.

4.   Membership --The subcommittee continues efforts to recruit new members to ensure a broad representation of federal agencies.

5.   Forum --The subcommittee continued as an information exchange and met when members expressed a need for information exchange on a topic in scope.

Activities Planned for CY 2003:

1. Provide a forum for interagency information exchanges in the areas of clearance, authorized release of solid materials with very low levels of associated radioactivity, orphan source identification and control, and import and export issues.

2. Continue to recruit members from agencies that could be impacted by regulations in the areas addressed by this subcommittee.

Risk Harmonization Subcommittee Highlights

Subcommittee Chair: Edward Regnier, DOE (202-586-5027)

The Risk Harmonization Subcommittee was initiated to review similarities and differences between NRC and EPA risk assessment and risk management approaches.

Accomplishments in CY 2002:

1. <u>Forum for Information Exchange</u>- The subcommittee provided a forum for information exchange and coordination between the agencies on issues and events related to risk assessment and risk management.

2. <u>Institutional Controls</u> - The subcommittee studied the use of institutional controls as part of the overall protective approach for disposal of waste. The requirements of NRC's regulation governing the disposal of high-level radioactive waste at Yucca Mountain, 10 CFR Part 63, were added to previously developed tables describing institutional control requirements in statutes and regulations. The subcommittee worked on a narrative description of these tables and of issues concerning the use of institutional controls.

3. <u>Literature Review</u> - The subcommittee reviewed and evaluated a report titled: Chemical and Radiation Environmental Risk Management at the Crossroads: Case Studies, which was written by the Environmental Law Institute and Johns Hopkins University.

Activities Planned for CY 2003:

1.   Provide a continuing forum for information exchange between the agencies on matters related to risk assessment and risk management.

2.   Incorporate the revised EPA regulations for disposal of HLW at Yucca Mountain, 40 CFR 197, into the institutional control tables. Complete a narrative description of these tables.

3.   Develop a set of ISCORS principles on the use of institutional controls which will represent agreed upon basic policies on the need for and the appropriate use and implementation of institutional controls.

4.   Read and discuss Determining Cleanup Goals at Radioactively Contaminated Sites: Case Studies, by the Interstate Technology and Regulatory Council and summarize the report for ISCORS.

5.   Hold, in conjunction with the clean up subcommittee, a workshop on land use scenarios for risk assessments for release of land following remediation.

6.   Monitor national and international organization activities related to risk harmonization and inform the main ISCORS Committee.

Possible Additional Projects for 2003:

7.   Prepare a white paper discussing issues and philosophies concerning the use of institutional controls

8.   Review the requirements in statutes, Executive orders, and OMB guidance related to the use of optimization methods (e.g., cost-benefit analysis, ALARA analysis, multi-attribute analysis) to support regulatory decision making. Develop tables which describe the statutory requirements for and conditions on the use of optimization.

Sewage Sludge Subcommittee Highlights

Subcommittee Co-chairs: Bob Bastian, EPA (202-260-7378)
Rosemary Hogan, NRC (301-415-7484)

The Sewage Sludge Subcommittee is conducting an NRC/EPA joint survey to collect information concerning radioactive materials in sewage sludge and ash from sewage treatment plants [referred to in industry as publicly owned treatment works (POTWs)], conduct dose modeling to help with the interpretation of the results of the survey, and to develop final guidance on this subject for the POTW owners and operators. This survey is being conducted to respond to recommendations from the General Accounting Office to 1) determine the extent to which radioactive contamination of sewage sludge and ash is occurring; 2) notify the POTWs that receive discharges from NRC's and Agreement State's licensees of the potential for radioactive concentration in the sludge and the possibility that they may need to test or monitor their sludge for radioactive content; and 3) establish limits for radioactivity in sludge and ash to ensure the health of treatment plant workers and the public.

Accomplishments in CY 2002:

1. Sewage Sludge Survey Results and Analysis -The sample analysis phase of the survey was completed and the final laboratory results were provided to the Subcommittee in October 2001 and the first draft of a "Joint NRC/EPA Sewage Sludge and Ash Radiological Survey: Survey Results and Analysis" report was provided to ISCORS in October 2001. The basic analytical results were presented in summary form at the July 2002 ISCORS public meeting and posted on the ISCORS website. ISCORS will be provided a final survey report document by the end of January 2003.

2. Dose Assessment - Feedback from ISCORS and others (including AMSA, which the Subcommittee met with to learn of the status of the National Biosolids Partnership's independent dose modeling activities) on the revised draft dose assessment document (posted on the ISCORS website for review in November 2001) was used to help further revise the document. Additional RESRAD runs were conducted by Argonne to address doses with and without Radon. The general dose assessment approach used in this effort was incorporated into a poster presentation at the American Geophysical Union meetings in May 2002 and a presentation at the Society for Risk Analysis meeting in December 2002. A draft final dose assessment document is targeted to be ready for ISCORS review in early 2003. The levels of radioactivity detected in sewage sludge and ash in the ISCORS survey and the dose assessment of those levels indicate that at most POTWs, radiation exposure to workers or to the general public is not likely to be a concern.

3. Guidance Document - The Subcommittee has utilized the results of the survey and dose modeling efforts and comments received in response to the June 2000 draft guidance document (posted on the ISCORS website) to develop a draft final guidance document targeted to be ready for ISCORS review in early 2003.

Activities Planned for CY 2003:

Activities Planned for CY 2003:

1.   Present the draft final survey report, dose modeling document, and guidance document to ISCORS for approval to post on the ISCORS website for public comment.

2.   Issue the final Survey Results and Analysis report.

3.   Submit the draft final dose modeling and guidance documents to the NRC's ACNW for formal peer review.

4.   Issue the final Dose Assessment report document.

5.   Issue the final Guidance Document for POTWs.

6.   Sunset the Sewage Sludge Subcommittee.

NORM Subcommittee Highlights

Subcommittee Chair: Loren Setlow, EPA (202-564-9445)

The NORM Subcommittee mission is to ensure effective communication and coordination among member agencies involved with regulatory, oversight, and disposal issues for naturally-occurring radioactive materials (NORM) and products containing NORM.

Accomplishments in CY 2002:

1. Held one meeting during this year, with members from seven Federal and one State agency.

2. Provided an opportunity for exchanges of information and updates for ongoing member agency programs, plus NRC's Jurisdictional Working Group.

3. Provided updates on status of proposed draft Part N model regulation for TENORM from the CRCPD, and Health Physics Society ANSI draft standard for TENORM.

4. Provided updates and information regarding upcoming and recently held international meetings on NORM/TENORM, including International Atomic Energy Agency technical meetings and working groups.

Activities Planned for CY 2003:

1. Complete the white paper on agency responsibilities for TENORM.

2. Provide input, as appropriate, to ongoing projects of the ISCORS Federal Guidance Subcommittee, and reviews of reports by the Sewage Sludge Subcommittee.

3. Review and comment on EPA's draft Technical Report on Uranium Mining TENORM when available.

4. Continue to receive updates on the CRCPD Part N model regulation, and ANSI draft standard for TENORM.

5. Provide comments on draft IAEA and other international reports on NORM/TENORM obtained by member agencies.

Federal Guidance Subcommittee Highlights

Subcommittee Co-Chairs: Michael Boyd, EPA (202) 564-9395
Harold Peterson, DOE (202) 586-9640

This subcommittee works with the EPA to produce Federal Guidance, including Presidential Guidance and technical reports, that supports the development of consistent national radiation protection standards and implementing guidance. The subcommittee serves as the forum for coordination among federal and state government agencies interested in any or all aspects of Federal Guidance. The forum ensures that all agencies have opportunities for input before and during development of Federal Guidance.

Accomplishments in CY 2003:

1. Protective Action Guides (PAGs) - The subcommittee was prepared to review any updates to the Protective Action Guides (PAGs) for possible publication as a Federal Guidance Technical Report. No updates were completed in 2002.

2. Federal Guidance for the General Public (FGGP) - This updated guidance, which would be signed by the President, advises federal agencies on protection of the general public from exposure to ionizing radiation. The current guidance dates back to the Eisenhower administration and was signed in 1960. The major effort during 2002 centered on efforts to incorporate interagency comments into a draft approved by an EPA intra-agency work group. It was decided by the ISCORS principals to move forward with a draft proposal that asks the public to comment on the major unresolved issues with the FGGP. At the close of 2002, this version was being reviewed by EPA's senior management for possible publication as a proposal in the Federal Register. EPA anticipates an active stakeholder effort in 2003 to solicit input for the final version.

3. Risk-Dose Methodology - The subcommittee presented ISCORS with a final version of a short methodology for estimating the cancer risk from a known radiation dose (total effective dose equivalent) when radionuclide-specific information is missing. ISCORS approved publication of this document as an ISCORS Technical Report at the September 18, 2002, mid-call.

4. White Paper on New Dosimetric Methods - The subcommittee delayed progress on this effort while pursuing completion of the Federal Guidance for the General Public. The subcommittee expects to give high priority to completion of this task early in 2003. Early in 2002, the subcommittee met with Dr. Keith Eckerman of Oak Ridge National Laboratory to go over his technical analysis, "Dosimetric Significance of the ICRP's Updated Recommendations, 1989-present: Implications for Federal Guidance for Occupational Exposure." This analysis will help support the completion of the White Paper.

5. Ecorisk - The subcommittee continued to support an informal working group (the ECORAD-WG) to focus its discussions concerning methods and guidance development regarding the evaluation of radiation as a stressor to the environment. Individual agencies within ISCORS have been actively developing guidance and approaches for evaluating doses to biota, and some of these activities are being coordinated for the subcommittee through this working group. Members of the workgroup attended and presented papers at the Third

International Symposium on the Protection of the Environment from Ionising Radiation (SPEIR 3) in Darwin, Australia, 22-26 July, 2002. Several agencies (EPA, NRC, and DOE) are collaborating on a pilot project to develop a RESRAD-BIOTA code for demonstrating protection of the environment from potential radiological impacts. NRC, DOE, and EPA coordinated their respective comments to the ICRP on a draft document on Protection of Biota which were due to ICRP in December, 2002.

Activities Planned for CY 2003:

1. Prepare a written discussion of the issues involved in changing from an ICRP Publication 26-based dosimetry system in the U.S. to one based on more recent or future recommendations of the ICRP. The discussion should explore in a qualitative manner the steps necessary to adopt new science across Agencies and investigate approaches for accommodating up-to-date science in assessing dose.

2. Support EPA as they analyze and respond to public comments and resolve any outstanding Agency concerns prior to seeking OMB and White House approval for issuing final Federal Guidance for the General Public in 2003.

3. Continue using the subcommittee as a forum for discussing issues and reviewing material related to protection of the environment from radioactivity. Keep ISCORS informed of developments in this area.

ISCORS MEMBER LIST

EPA

Ms. Elizabeth Cotsworth, Director
Office of Radiation and Indoor Air (6608J)
U.S. Environmental Protection Agency
1200 Pennsylvania Avenue, NW
Washington, D.C. 20460

202-564-9320
cotsworth.elizabeth@epa.gov

Mr. Frank Marcinowski, Director (ISCORS Co-chair)
Radiation Protection Division
Office of Radiation and Indoor Air (6608J)
U.S. Environmental Protection Agency
1200 Pennsylvania Avenue, NW
Washington, D.C. 20460

202-564-9290
fax: 202-565-2065
marcinowski.frank@epa.gov

Ms. Kathryn Klawiter (ISCORS coordinator)
Radiation Protection Division
Office of Radiation and Indoor Air (6608J)
1200 Pennsylvania Avenue, NW
Washington, D.C. 20460

202-564-9228
fax: 202-565-2042
klawiter.kathryn@epa.gov

NRC

Dr. Carl Paperiello, Deputy Executive Director for Materials,
Research, and State Programs
Office of the Executive Director for Operations, Mail Stop O-16E15
U.S. Nuclear Regulatory Commission
Washington, D.C. 20555

301-415-1705
cjp1@nrc.gov

Mr. John T. Greeves, Director (ISCORS Co-chair)
Division of Waste Management (DWM), Mail Stop T-7J8
Office of Nuclear Material Safety and Safeguards
U.S. Nuclear Regulatory Commission
Washington, D.C. 20555

301-415-7437
fax: 301-415-5397
jtg1@nrc.gov

Mr. James Kennedy (ISCORS coordinator)
Division of Waste Management (DWM), Mail Stop T-7J8
Office of Nuclear Material Safety and Safeguards
U.S. Nuclear Regulatory Commission
Washington, D.C. 20555

301-415-6668
fax: 301-415-5397
jek1@nrc.gov

DOD

Mr. D. Michael Schaeffer/TDND
Defense Threat Reduction Agency
8725 John J, Kingman Road, Stop 6201
Ft. Belvoir, VA 22060-6201

703-325-2407
fax: 703-325-2951
dennis.schaeffer@dtra.mil

CAPT David Farrand, USN, MSC (alternate)
Head, Radiological Controls and Health Branch
Office of the Chief of Naval Operations (N455)
2211 South Clark Place, Room 680
Arlington, VA 22202-3735

703-602-5365
fax: 703-602-4786
farrand.david@hq.navy.mil

DOE

Mr. Andrew Wallo, EH-412, Director
Air, Water, and Radiation Division
U.S. Department of Energy
Washington, D.C. 20585

202-586-4996
fax: 202-586-3915
andrew.wallo@eh.doe.gov

Dr. Colleen Ostrowski (ISCORS Coordinator)
Air, Water and Radiation Division
U.S. Department of Energy
Washington, D.C. 20585

202-586-4997
fax: 202-586-3915
colleen.ostrowski@eh.doe.gov

DOL/OSHA

Mr. Steven Witt, Director
Health Standards Programs
Occupational Safety and Health Administration
U.S. Department of Labor
Room N-3718
200 Constitution Avenue, N.W.
Washington, DC 20210

202-693-1950
fax: 202-693-1678
steven.witt@osha.gov

Dr. Chia Chen, Senior Industrial Hygienist
Occupational Safety and Health Administration
U.S. Department of Labor
Room N-3718
200 Constitution Avenue, N.W.
Washington, D.C. 20210

202-693-2266
fax: 202-693-1678
chia.chen@osha-no.osha.gov

DOT

Mr. Robert A. McGuire, Associate Administrator
Office of Hazardous Materials Safety, DHM-1
U.S. Department of Transportation
400 Seventh Street, S.W.
Washington, D.C. 20590

202-366-0656
fax: 202-366-5713
robert.mcguire@rspa.dot.gov

Mr. Rick Boyle, Chief
Radioactive Materials Branch, DHM-23
U.S. Department of Transportation
400 Seventh Street, S.W.
Washington, D.C. 20590

202-366-4545
fax: 202-366-3753
rick.boyle@rspa.dot.gov

Mr. Fred Ferate
Radioactive Materials Branch, DHM-23
U.S. Department of Transportation
400 Seventh Street, S.W.
Washington, D.C. 20590

202-366-4498
fax: 202-366-3753
fred.ferate@rspa.dot.gov

DHHS

DHHS

John L. McCrohan, Director
Division of Mammography Quality and Radiation Programs
Office of Health and Industry Programs
Center for Devices and Radiological Health
Food and Drug Administration
1350 Piccard Drive [Room 220Q]
Rockville MD 20850

301-594-3332
fax: 301-594-3306
jlm@cdrh.fda.gov

OMB

Mr. Bryon Allen
Natural Resources Branch
Office of Information and Regulatory Affairs
Room 10202
New Executive Office Building
725 17th Street
Washington, DC 20503

202-395-3087
fax: 202-395-7285
bryon_p._allen@omb.eop.gov

OBSERVER MEMBERS

Office of Science and Technology Policy

Mr. Robert Marianelli
Office of Science and Technology Policy
17th & Pennsylvania Avenue, N.W.
Washington, DC 20500

202-456-6134
fax: 202-456-6027
rmariane@ostp.eop.gov

Dr. Miriam Forman
Office of Science and Technology Policy
17th & Pennsylvania Avenue, N.W.
Washington, DC 20500

202-456-6134
fax: 202-456-6027
mforman@ostp.eop.gov

State Observers

Mr. Steve Collins
Illinois Department of Nuclear Safety
Office of Radiation Safety
1035 Outer Park Drive
Springfield, IL 62704

217-785-6982
fax: 217-782-1328
collins@idns.state.il.us

Mr. David J. Allard, CHP, Bureau Director
Bureau of Radiation Protection
Pennsylvania Department of Environmental Protection
Rachel Carson State Office Building
P.O. Box 8469
Harrisburg, PA 17105-8469

717-787-2480
fax: 717-783-8965
dallard@state.pa.us

SUBCOMMITTEE MEMBER LIST

Subcommittee members conduct reviews on specific issues to bring to the full ISCORS membership for consideration and decisions.

CLEANUP SUBCOMMITTEE

ORGANIZATIONS	MEMBERS	PHONE NUMBER
EPA	Tony Wolbarst, Chair	202-564-9392
	Weihsueh Chiu	
	Bonnie Gitlin	
	Ben Hull	
	Stuart Walker	
NRC	Ralph Cady	
	Boby Abu-Eid	
	Sandra Wassler	
DOD	Julie Peterson	
	Richard Wright	
DOE	Stephen Domotor	
States	Debra McBaugh, Washington	

MIXED WASTE SUBCOMMITTEE

ORGANIZATIONS	MEMBERS	PHONE NUMBER
DOE	Gus Vazquez, Chair	202-586-7629
	Jim Antizzo	
	William Fortune	
	Ed Regnier	
	Andy Wallo	
EPA	Nancy Hunt	
	David Levenstein	
	Dan Schultheisz	
NRC	Suzanne Woods	
DOD	Kelly Crooks	
DOT	Fred Ferate	
States	Paul Merges, New York	

RECYCLE SUBCOMMITTEE

ORGANIZATIONS	MEMBERS	PHONE NUMBER
NRC	Robert Meck, Chair	301-415-6205
	Doug Broaddus	
	Elaine Brummett	
	Frank Cardile	
	Carl Feldman	
	Giorgio Gnugnoli	
	Tony Huffert	
EPA	Deborah Kopsick	
	Mary Clark	
DOD	CAPT Dave Farrand	
DOE	Gus Vazquez	
	Lee Bishop	
	John Neave	
DOL/OSHA	Chia Chen	
States	Steve Collins, Illinois	

RISK HARMONIZATION SUBCOMMITTEE

ORGANIZATIONS	MEMBERS	PHONE NUMBER
DOE	Edward Regnier, Chair	202-586-5027
	Colleen Ostrowski	
EPA	Mike Boyd	
	Ernesto Brown	
	David Pawel	
	Stuart Walker	
NRC	Dennis Damon	
	James Kennedy	
DOD	Michael Schaeffer	
DOL/OSHA	Chia Chen	
DOT	Fred Ferate	
States	Jill Lipoti, New Jersey	

Institutional Controls Workgroup

NRC	James Kennedy	
EPA	Ernesto Brown	
DOE	Colleen Ostrowski	

Optimization Workgroup

NRC	Dennis Damon	
EPA	Mike Boyd	
DOE	Ed Regnier	

SEWAGE SLUDGE SUBCOMMITTEE

ORGANIZATIONS	MEMBERS	PHONE NUMBER
EPA	Bob Bastian, Co-Chair	202-260-7378
	Weihsueh Chiu	
	Mark Doehnert	
	Alan Rubin	
	David Saunders	
	Loren Setlow	
	Behram Shroff	
	Anthony Wolbarst	
NRC	Rosemary Hogan, Co-Chair	301-415-7484
	Lydia Chang	
	Giorgio Gnugnoli	
	Andrea Jones	
	Tom Nicholson	
	Stephen Salomon	
	Duane Schmidt	
DOE	James Bachmaier	
	Judy Odoulamy	
	Harold Peterson	
States	Jenny Goodman, New Jersey	
Local Government	Kevin Aiello, Middlesex County Utilities Authority	
	Tom Lenhart, Northeast Ohio Regional Sewer District	

NORM SUBCOMMITTEE

ORGANIZATIONS	MEMBERS	PHONE NUMBER
EPA	Loren Setlow, Chair	202-564-9445
NRC	James Kennedy	
DOE	Bill Hochheiser	
	Alexander Williams	
DOT	Fred Ferate	
OSHA/DOL	Chia Chen	
DOD	CMDR Bill Adams	
	Ramchandra Bhat	
	CAPT David Pugh	
HHS	CMDR Shawn Googins	
States	Walter Cofer, Florida	

FEDERAL GUIDANCE SUBCOMMITTEE

ORGANIZATIONS	MEMBERS	PHONE NUMBER
EPA	Mike Boyd, Co-Chair	202-564-9395
	Keith Matthews	
	Scott Monroe	
	Phillip Newkirk	
	Jerry Puskin	
	Lowell Ralston	
	Loren Setlow	
	Stuart Walker	
DOE	Hal Peterson, Co-Chair	202-586-9640
	Steve Domotor	
	Andy Wallo	
NRC	Roger Pedersen	
	Sami Sherbini	
	Vince Holahan	
DOT	Fred Ferate	
HHS	Sam Keith	
OSHA/DOL	Chia Chen	
DOD	Mike Schaeffer	
States	John Erickson, Washington	

APPENDIX A
ISCORS Charter

CHARTER FOR
INTERAGENCY STEERING COMMITTEE ON RADIATION STANDARDS

Purpose of Committee

The purpose of this committee is to foster early resolution and coordination of regulatory issues associated with radiation standards.

Membership

1. Agencies represented on the committee include the U.S. Environmental Protection Agency, U.S. Nuclear Regulatory Commission, U.S. Department of Energy, U.S. Department of Defense, U.S. Department of Transportation, the Occupational Safety and Health Administration of the U.S. Department of Labor, the U.S. Department of Health and Human Services, and any successor agencies.

2. The Office of Science and Technology Policy and the Office of Management and Budget will be invited as observers at meetings because of their responsibilities for coordination of science policy and regulation policy, respectively.

3. The committee will be co-chaired by the EPA and NRC representatives for the first two years, after which the committee will determine whether the chairmanship should be rotated among additional agencies, or whether the chairmanship should be held by a single agency.

4. Other departments and agencies will be invited to participate in forming consensus for specific issues as voting members when their interests and responsibilities are involved.

Objectives

The objectives of the committee include the following:

1. Facilitate a consensus on allowable levels of radiation risk to the public and workers.

2. Promote consistent and scientifically sound risk assessment and risk management approaches in setting and implementing standards for occupational and public protection from ionizing radiation.

3. Promote completeness and coherence of Federal standards for radiation protection.

4. Identify interagency radiation protection issues and coordinate their resolution.

Implementation

The committee will conduct its activities in accordance with the attached understandings and procedures.

UNDERSTANDINGS AND PROCEDURES FOR
THE INTERAGENCY STEERING COMMITTEE ON RADIATION STANDARDS

Participation

1. Various offices and agencies within each agency may participate in the committee meetings. Each agency will develop a unified position and present that position at committee meetings. Each agency representative is responsible for developing their coordinated agency position in preparation for reaching committee consensus.

2. Agencies will be represented at the meetings by senior level, career government employees, who are engaged in policy matters for the agency.

3. Official agency representatives will be identified in writing to the co-chairpersons by the Assistant Administrator, Secretary, or Commissioner, as appropriate.

4. Committee meetings involve pre-decisional intragovernmental discussions and, as such, are not open for observation by members of the public or media.

5. The committee may, from time to time, revise the charter based on the consensus views of the committee, including such items as membership, responsibility for chairing the committee, and objectives.

Decisions

1. The committee has not been delegated any authorities established by law, regulation, Executive Order, or other administrative mechanism to act in lieu of formal agency action. The objectives of the committee are described in the committee charter.

2. The committee will make every effort to base decisions on consensus. Consensus reflects acceptance among the voting agencies.

3. Each agency will have a single vote in reaching consensus on specific issues. If a consensus cannot be reached, committee recommendations will reflect the lack thereof and include the opportunity for agencies to attach minority views to any documentation of the recommendations.

4. Recommendations on specific issues will be provided to the heads of member agencies, OMB, and OSTP.

Meetings

1. Responsibility for hosting the meetings will rotate among the agencies. The host agency is responsible for developing a mutually agreeable meeting date and time, informing the agencies at least two weeks in advance of the

meeting date, distributing a draft agenda for the meeting, arranging for a meeting facility, and documenting and distributing summary meeting notes.

2. Summary meeting notes will be provided by the host agency to designated representatives of each of the member agencies, OMB, OSTP, and, as appropriate, Congressional contacts and other groups. The host agency will distribute draft notes within one week of the committee meeting and final meeting notes at least two weeks before the next committee meeting. NRC will also place a copy of the summary meeting notes in the Public Document Room.

3. The committee will establish a plan for approximately a six-month period. Specific agendas will be developed for each meeting based on the general plan.

4. The committee will meet approximately once each calendar quarter, with more frequent meetings, as needed to address specific issues.

Subcommittees

1. The committee may create subcommittees to focus on specific issues or activities (e.g., recycling criteria, risk harmonization, cleanup standards). Subcommittees will follow the same understandings and procedures as the full committee.

2. Subcommittees will meet at a frequency and location as determined necessary by the subcommittee.

FOR THE U.S. NUCLEAR REGULATORY COMMISSION,

Hugh L. Thompson   _12-18-95_
Hugh L. Thompson, Jr.   Date
Deputy Executive Director for
   Nuclear Materials Safety,
   Safeguards and Operations Support

FOR THE U.S. ENVIRONMENTAL PROTECTION AGENCY,

Mary Nichols   _1-16-96_
Mary Nichols   Date
Assistant Administrator
Office of Air and Radiation

FOR THE U.S. DEPARTMENT OF ENERGY,

Tara O'Toole   _2/4/96_
Tara O'Toole, M.D., M.P.H.   Date
Assistant Secretary for
   Environment, Safety and
   Health

FOR THE U.S. DEPARTMENT OF DEFENSE,

                                            JUL 1 2 1996
Paul Kaminski
Paul G. Kaminski   Date
Under Secretary of Defense
   (Acquisition & Technology)

FOR THE U.S. DEPARTMENT OF TRANSPORTATION,

                                            OCT 2 4 1996
D. K. Sharma
Dr. Dharmendra K. Sharma   Date
Administrator
Research and Special Programs
   Administration

FOR THE U.S. DEPARTMENT OF LABOR,

Greg Watchman   _3/21/97_
Greg Watchman   Date
Acting Assistant Secretary for
   Occupational Safety and
   Health Administration

FOR THE US DEPARTMENT OF HEALTH AND HUMAN SERVICES

_____

Jo Ivey Boufford, M.D.
Acting Assistant Secretary for Health

5 - 16 - 97

Date

APPENDIX B
Glossary

ACRONYM LIST

ACNW	Advisory Committee on Nuclear Waste
ALARA	As low as is reasonably achievable (radiation dose)
AMSA/WEF	American Metropolitan Sewage Association/ Water Environmental Federation
ANSI	American National Standards Institute
CERCLA	Comprehensive Environmental Response, Compensation, and Liability Act
CIRRPC	Committee for Interagency Radiation Research and Policy Coordination
CRCPD	Conference of Radiation Control Program Directors
DHHS	Department of Health and Human Services
DOE	Department of Energy
DOL	Department of Labor
DOT	Department of Transportation
ELI	Environmental Law Institute
EPA	Environmental Protection Agency
FGGP	Federal Guidance for the General Public
IAEA	International Atomic Energy Agency
ICRP	International Commission on Radiation Protection
ISCORS	Interagency Steering Committee on Radiation Standards
NAS	National Academy of Science
NCRP	National Council on Radiation Protection and Measurements
NORM	Naturally Occurring Radioactive Material
NRC	Nuclear Regulatory Commission
OMB	Office of Management and Budget
OSHA	Occupational Safety and Health Administration
OSTP	Office of Science and Technology Policy

PAG	Protective Action Guide
POTW	Publicly-Owned Treatment Works
RCC	Radiological Control Criteria
RCRA	Resources Conservation and Recovery Act
TENORM	Technologically Enhanced Naturally Occurring Radioactive Material

Radiation Unit Abbreviations

mrem/y	millirem/year
Sv/a	Sievert per annum

NRC FORM 335
(2-89)
NRCM 1102,
3201, 3202

U.S. NUCLEAR REGULATORY COMMISSION

BIBLIOGRAPHIC DATA SHEET

(See instructions on the reverse)

1. REPORT NUMBER
(Assigned by NRC, Add Vol., Supp., Rev., and Addendum Numbers, if any.)

NUREG-1707, Volume 5
ISCORS-03-01

2. TITLE AND SUBTITLE

Interagency Steering Committee on Radiation Standards, 2002 Annual Report

3.	DATE REPORT PUBLISHED	
	MONTH	YEAR
	May	2003

4. FIN OR GRANT NUMBER

N/A

5. AUTHOR(S)

Kathryn A. Klawiter, U.S. Environmental Protection Agency
James E. Kennedy, U.S. Nuclear Regulatory Commission

6. TYPE OF REPORT

Final

7. PERIOD COVERED *(Inclusive Dates)*

01/01/2002 - 12/31/2002

8. PERFORMING ORGANIZATION - NAME AND ADDRESS *(If NRC, provide Division, Office or Region, U.S. Nuclear Regulatory Commission, and mailing address; if contractor, provide name and mailing address.)*

Division of Waste Management, Office of Nuclear Material Safety and Safeguards, U.S. Nuclear Regulatory Commission, Washington DC 20555

9. SPONSORING ORGANIZATION - NAME AND ADDRESS *(If NRC, type "Same as above"; if contractor, provide NRC Division, Office or Region, U.S. Nuclear Regulatory Commission, and mailing address.)*

Same as 8 above

10. SUPPLEMENTARY NOTES

.

11. ABSTRACT *(200 words or less)*

The Interagency Steering Committee on Radiation Standards (ISCORS) prepared this annual report for ISCORS member agencies to document ISCORS' year 2002 activities and plans for 2003. The report identifies both past accomplishments and goals for the future. The objectives of ISCORS include: (1) facilitating a consensus on acceptable levels of radiation risk to the public and workers; (2) promoting consistent risk assessment and risk management approaches in setting and implementing standards for occupational and public protection from ionizing radiation; (3) promoting completeness and coherence of Federal standards for radiation protection; and (4) identifying interagency issues and coordinating their resolution.

12. KEY WORDS/DESCRIPTORS *(List words or phrases that will assist researchers in locating the report.)*

ISCORS
risk management
risk assessment
standards

13. AVAILABILITY STATEMENT

unlimited

14. SECURITY CLASSIFICATION

(This Page)

unclassified

(This Report)

unclassified

15. NUMBER OF PAGES

16. PRICE

NRC FORM 335 (2-89)

This form was electronically produced by Elite Federal Forms, Inc.

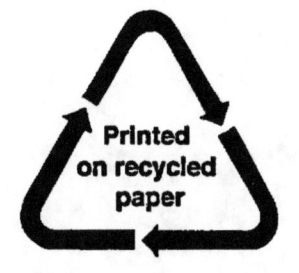

Federal Recycling Program